SPACE WONDERS

OUTER PLANETS

BLACK RABBIT BOOKS

TABLE OF CONTENTS

1. Far from the Sun 4

2. Jupiter 6

3. Saturn 10

4. Uranus 13

5. Neptune 16

6. Strange Worlds 19

More to Explore 22

1

Far from the Sun

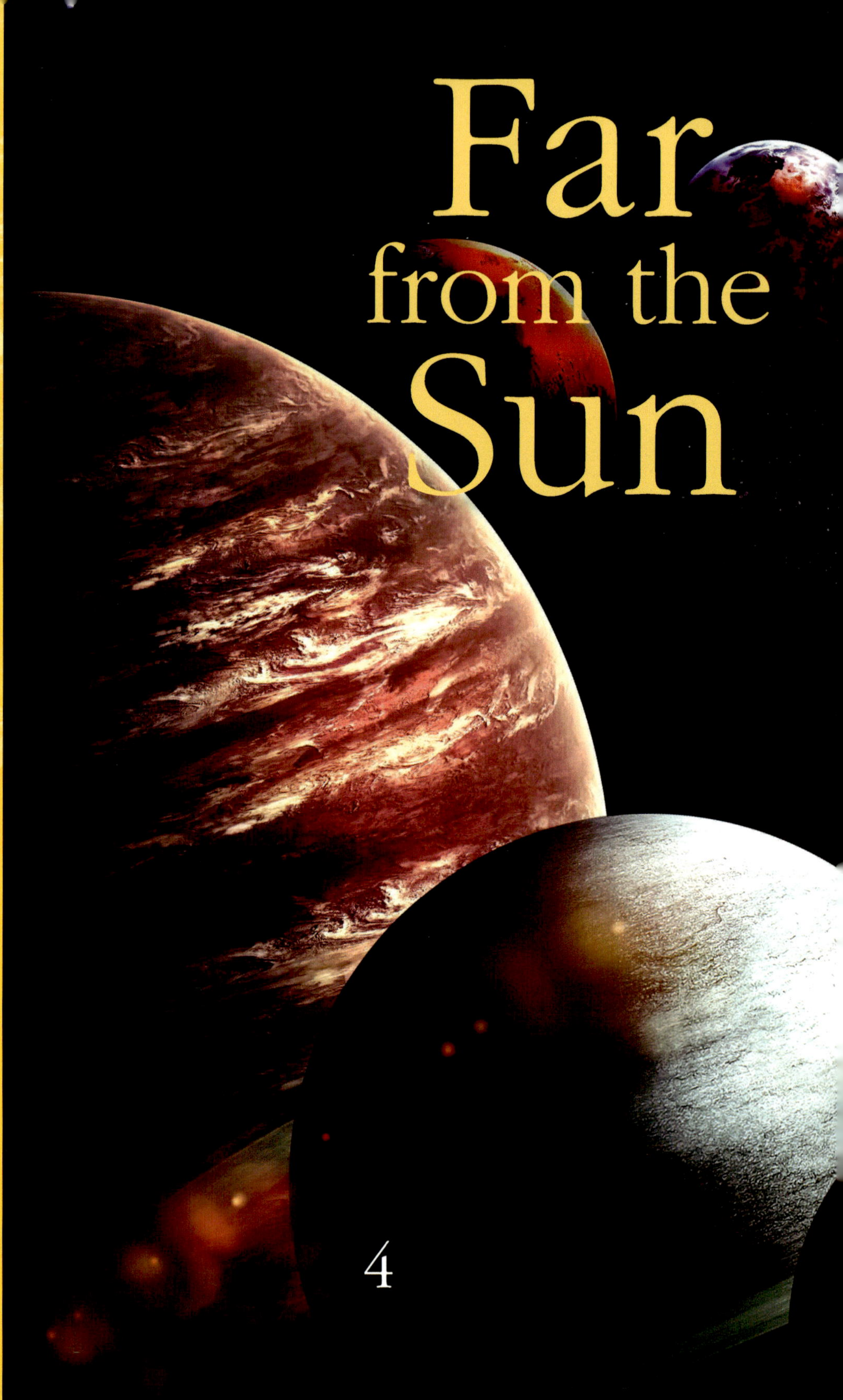

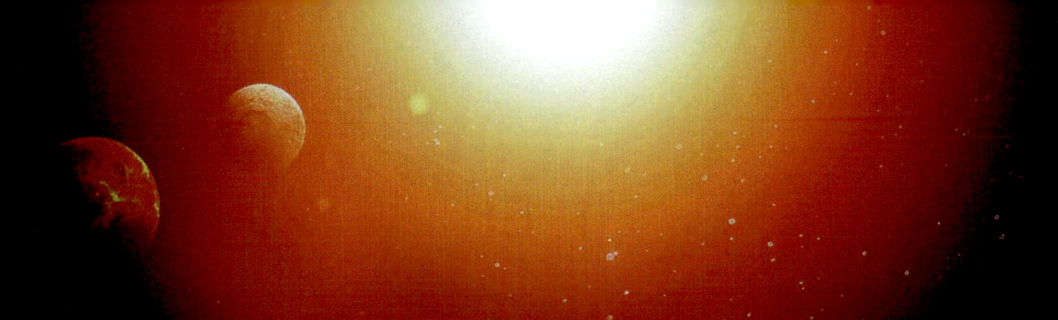

There are eight planets in our **solar system**. Four are closest to the sun. They are called the inner planets. The other four are much farther away. They are the outer planets. And they are very different.

The outer planets are called gas giants. They are much bigger than the inner planets. And they are mostly made of helium and hydrogen. These are gases.

Think About It
Why do you think the outer planets are different from the inner planets?

2

Juno

Jupiter

Jupiter is the largest planet in our solar system. It is so big that more than 1,300 Earths could fit inside it! This giant planet is a cold world. Its temperature is about -234 degrees Fahrenheit (-148 degrees Celsius). The **atmosphere** is made of thick clouds. There is no solid ground.

Juno is a space **probe**. It studies Jupiter. One area is called the Great Red Spot. It is a big storm. Winds blow more than 580 miles (933 kilometers) an hour! The storm has been going on for hundreds of years.

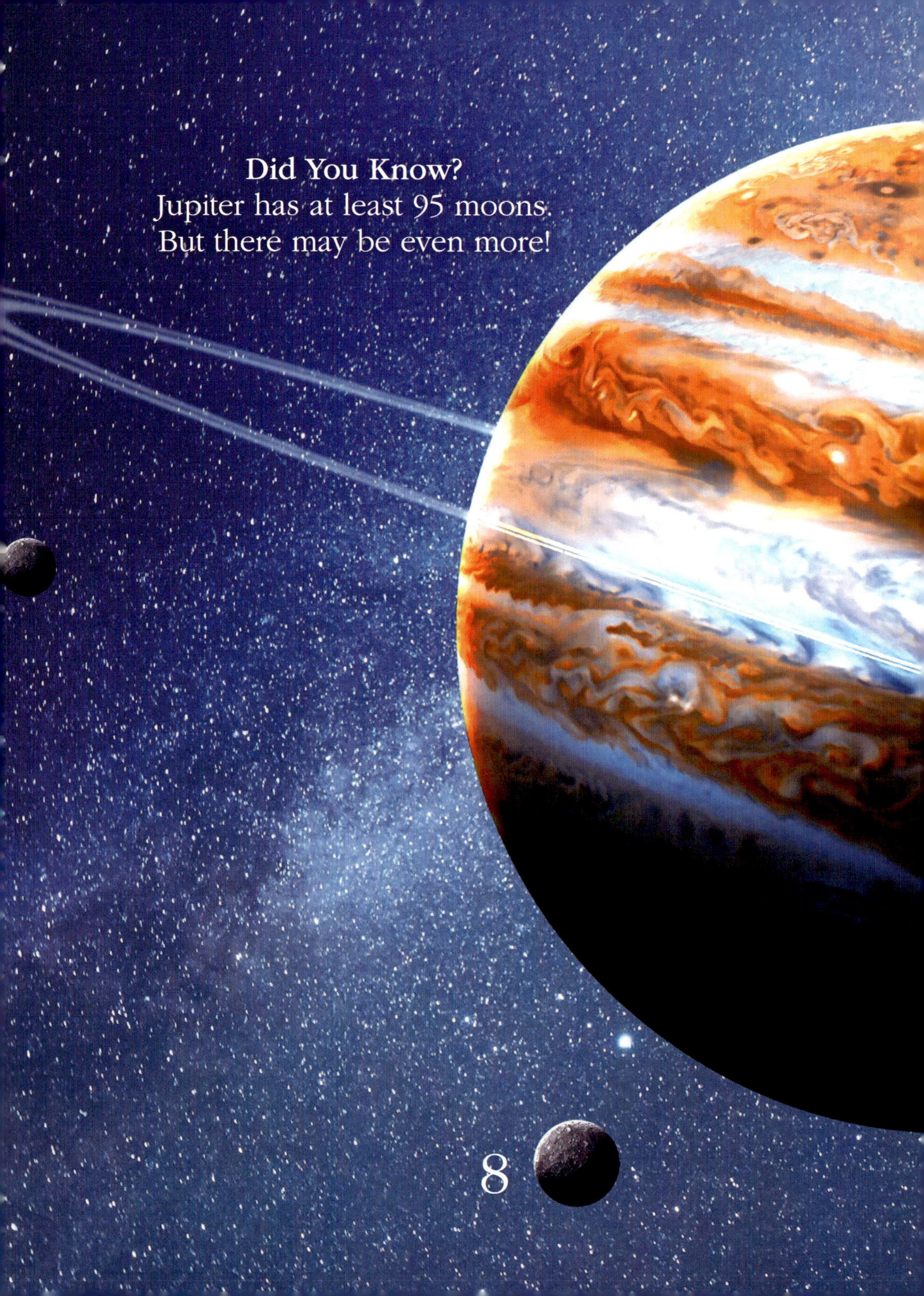

Did You Know?
Jupiter has at least 95 moons.
But there may be even more!

3 Saturn

Did You Know?
Saturn is less **dense** than water. It could float in a giant bathtub!

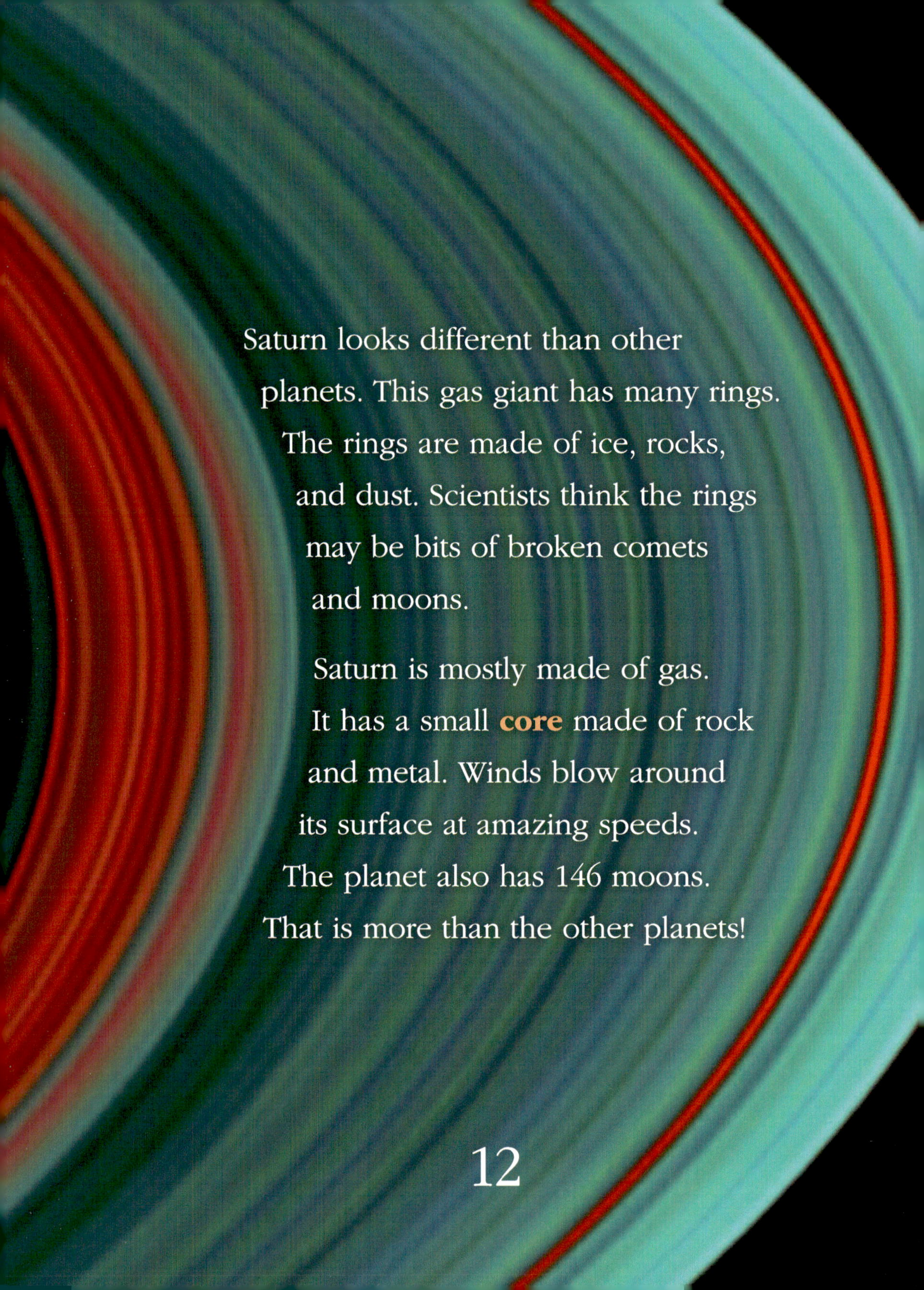

Saturn looks different than other planets. This gas giant has many rings. The rings are made of ice, rocks, and dust. Scientists think the rings may be bits of broken comets and moons.

Saturn is mostly made of gas. It has a small **core** made of rock and metal. Winds blow around its surface at amazing speeds. The planet also has 146 moons. That is more than the other planets!

4 Uranus

This planet is so cold, it is called an ice giant. Uranus is the solar system's coldest planet. Its temperature is a chilly -320°F (-195°C). Winds blow up to 560 miles (900 km) per hour.

Millions of years ago, something hit Uranus. The planet was knocked sideways. It lies on its side. It looks like a rolling ball. That means the planet's poles are all light or all dark for years and years.

Did You Know?
Uranus has two sets of rings. They are hard to see.

5

Neptune

Neptune is very far away. Its **orbit** takes 165 Earth years. This ice giant looks blue. The gases in its atmosphere absorb red light. It reflects the sun's blue light back into space.

In 1989, *Voyager 2* flew past Neptune. It is the only spacecraft to do this. It flew about 3,000 miles (4,800 km) above the planet's north pole. Today, space **telescopes** study it. They take pictures.

Think About It

How can we observe planets where spacecraft cannot land?

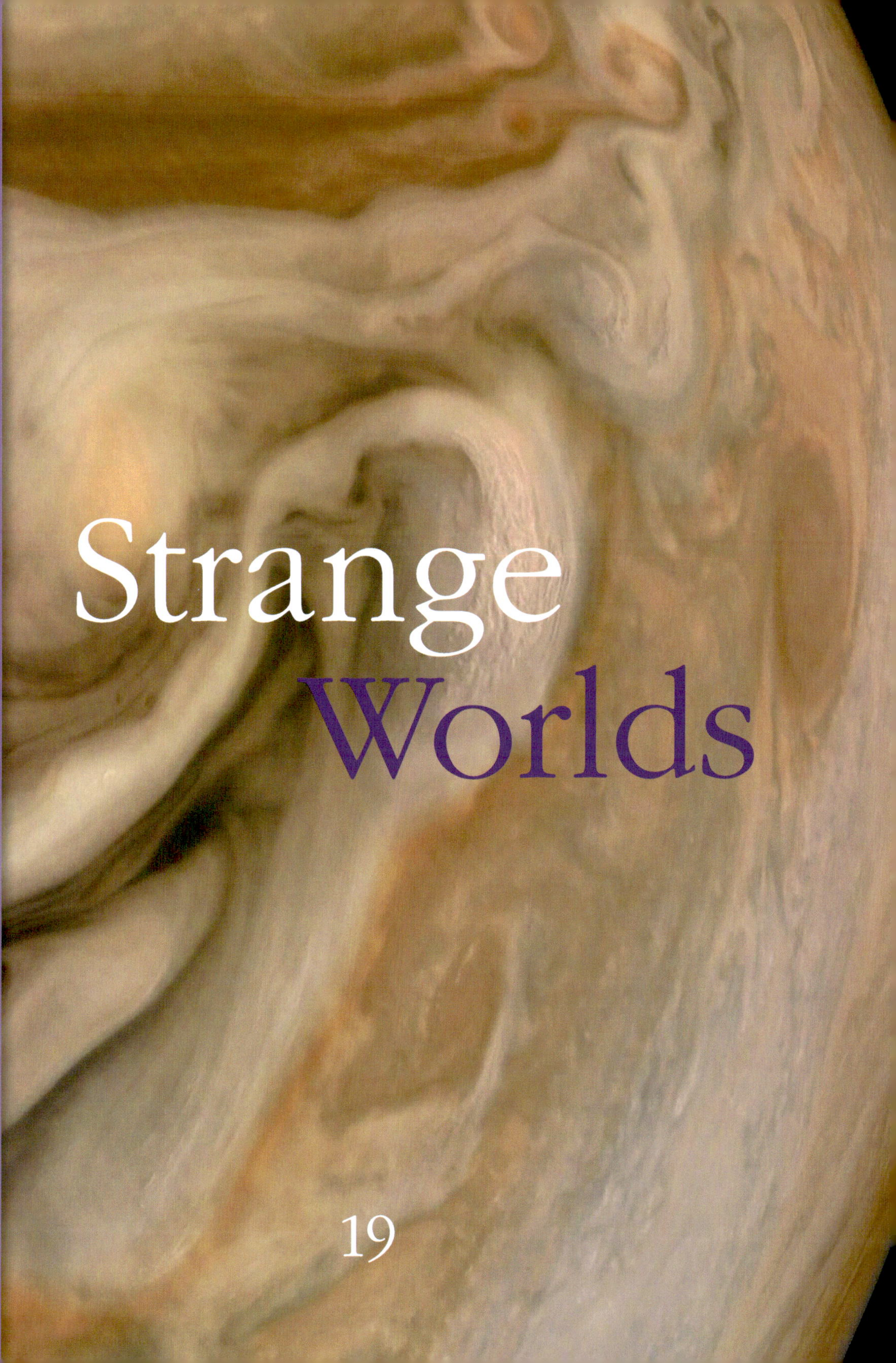

6

Strange Worlds

Scientists want to learn more about the outer planets. They send space probes. These study the planets. Telescopes also take pictures of these worlds.

It is hard to study objects that are so far away. But it is important to learn more about the outer planets. When we study them, we learn more about the solar system.

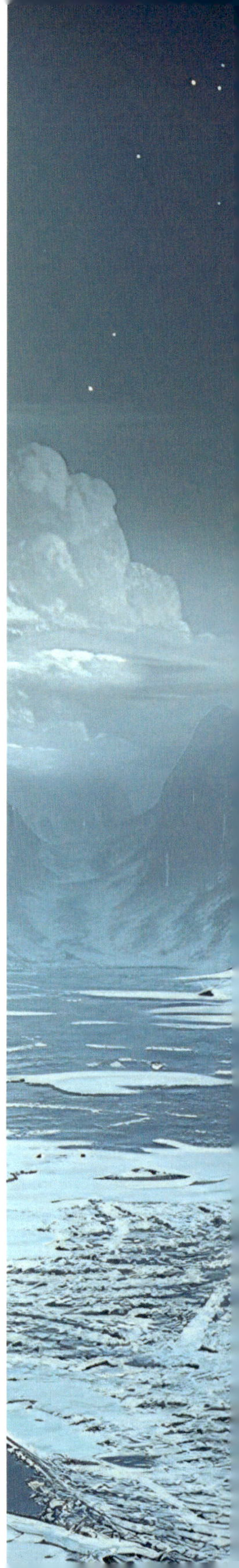

MORE TO EXPLORE

FANTASTIC FACTS

The pressure on Uranus is so strong, it would destroy any spacecraft.

Jupiter spins faster than any other planet. Its day is only 10 hours!

Pioneer 10 was the first spacecraft to fly past Jupiter. It made the trip in 1972.

Cassini arrived at Saturn in 2004. It studied the planet, its rings, and its moons for about 13 years.

Saturn is the farthest planet you can see without a telescope.

Scientists think a hot ocean might be hidden under Neptune's icy clouds.

MORE TO EXPLORE

COOL COMPARISONS

How many Earths could fit inside each outer planet?

MORE TO EXPLORE

RESOURCES

Glossary

atmosphere (AT-muhs-feer) The gases that surround a planet.

core (KOHR) The central part of a planet or other body.

dense (DENS) Heavier than most things of the same size.

orbit (AWR-bit) The path taken by one body circling around another body.

probe (PROHB) A device used to collect information from outer space and send it back to Earth.

solar system (SOH-luhr SYS-tum) The planets and space objects that circle the sun.

telescope (TEL-uh-skohp) A tube-shaped instrument that you look through to see things that are far away.

Read More

Leaf, Christina. *The Outer Planets*. Minneapolis: Bellwether Media, 2023.

Nelson, Louise. *The Solar System*. Minneapolis: Jump!, Inc., 2024.

Index

gases, 5, 12, 17
moons, 8, 12, 22
orbit, 14, 17
rings, 12, 14, 22
scientists, 12, 20, 22
size, 7, 23
spacecraft, 7, 18, 20, 22
temperature, 7, 14
weather, 7, 12, 14

TOP RANK is published by Black Rabbit Books, P.O. Box 227, Mankato, MN, 56002.

• Top Rank is an imprint of Black Rabbit Books • Series designed by Danny Nanos • Book designed by Jason Knudson • Photographs © Adobe Stock/Pavel, 14–15, visoot, 21, woopics, 23; Freepik/alexkoral, 13, wirestock, 2–3; NASA/JPL, 6–7, Voyager 2, 23;NASA/JPL-Caltech, 18; NASA/JPL-Caltech/SwRI/MSSS Image processing: Brian Swift, 19; NASA/JPL/Space Science Institute, 10–11; NASA/JPL/University of Colorado, 12; Shutterstock buradaki, cover, 16–17, Vadim Sadovski, 4–5, 8–9, 23; Wikimedia Commons/Martin Schweiger, MIT, 2, 23, NASA/JPL-Caltech/B. Jónsson, 13, 18 • Printed in the United States of America

Library of Congress Cataloging-in-Publication Data: Names: Mattern, Joanne, 1963- author. | Title: Outer planets / Joanne Mattern. | Description: Mankato, MN: Black Rabbit Books, [2026] | Series: Space wonders | Includes bibliographical references and index. | Audience: Ages 8–11 | Audience: Grades 2–3 | Identifiers: LCCN 2025011958 (print) | LCCN 2025011959 (ebook) | ISBN 9781645825128 (library binding) | ISBN 9781645825302 (paperback) | ISBN 9781645825487 (ebook) | Subjects: LCSH: Outer planets—Juvenile literature. | Classification: LCC QB659 .M37 2026 (print) | LCC QB659 (ebook) | DDC 523.4—dc23/eng/20250324 | LC record available at https://lccn.loc.gov/2025011958 | LC ebook record available at https://lccn.loc.gov/2025011959